By Donna Scott-Nusrala

Graphics Editor: Aaron Nusrala

For more information or to order the accompanying educator's metric kit, please contact:
Donna Scott-Nusrala
Telephone: 440-554-7505
Email Author at: donna@amopublishing.com
www.amopublishing.com

Library of Congress Control Number: 2010936571
ISBN 978-935268-77-2

Author: Donna Scott-Nusrala
Illustrations: Kevin Scott Collier
Graphics Editor: Aaron Nusrala

in conjunction with

Printed in the United States of America

This book is dedicated to…

- Patrick, Aaron, Dillon and Adam
- With a big thank you to Aaron for graphics editing
- Mom and Dad for “Sneezles” by A.A. Milne
- All of the children, families and staff of Euclid City Schools and Hawken School who blessed my life and inspired me.
- To Mary Haller and Dr. Jann Gallagher
- Dr. Frank Johns, Cleveland State University
- To Karen for guidance
- And to Courtney Strah

10
10

Take a look in this house with an address of 10.
There's a family that measures again and again.

They shrink, they grow, they morph and they fly!
The family is "Metric." I'll tell you why.

Granddaddy Kilo is the strongest.
He is the biggest and stretches the longest.

Hecto is dad, a rather large man,

but smaller than Kilo. That is the plan.

Deka is ten times slighter in size.

She is the mom, I would surmise.

She's followed by the family crest.
It shows which last name is the best.

When measuring volume, the crest morphs to Liter.
Mass names them Gram, and length makes them Meter.

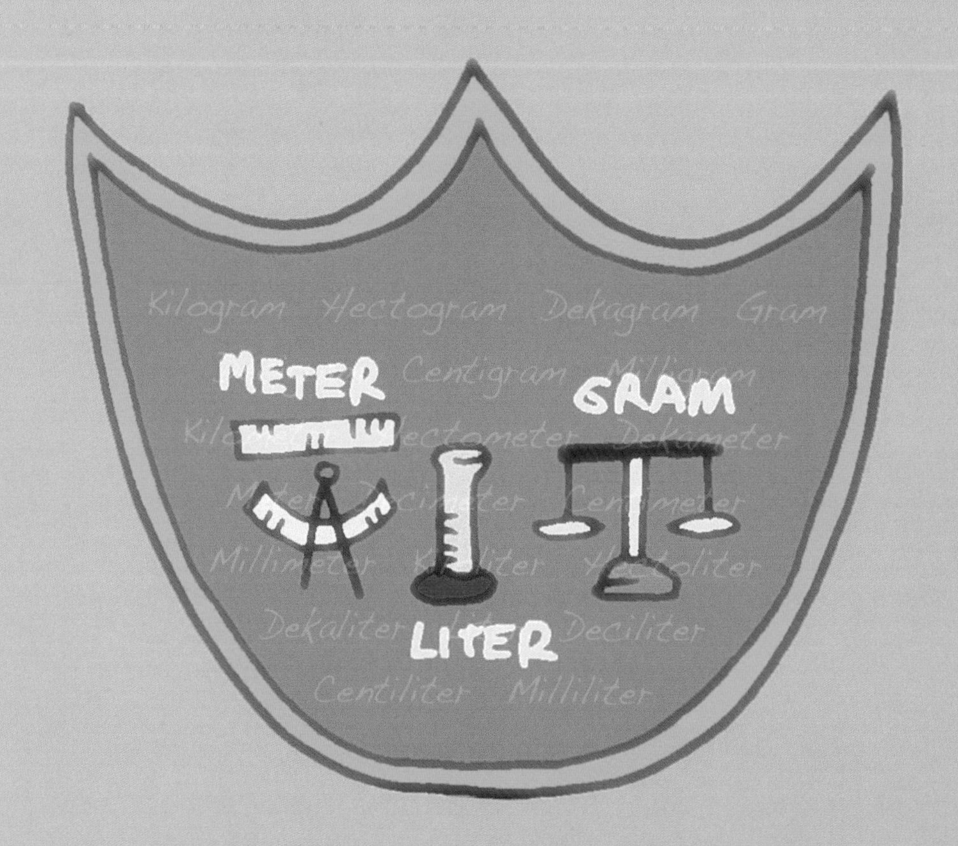

There's one more thing that they can do.
Each last name can measure too.

Deci, their daughter, is next in size.
She morphs rather small, when she tries.

And then there is Centi, their only son.
He thinks shrinking and morphing are fun.

Milli is silly. She gets so small.
You must look closely to see her at all.

LOOK at the crest!
It's giving a name...
It's growing to "Gram"...
"Mass!" they claim.

When their last name is Gram,
they won't fail.
They're measuring mass,
using a scale.

Kilogram is the largest of mass.
He is as much as a textbook in class.

Hectogram is lighter
by ten times as much.
The mass of an apple,
an orange, or such.

An egg or a cookie,
watch Dekagram change.
She's showing her mass.
Isn't she strange?

**One gram would be
as much as a dime.
You'd find that true,
if you took the time.**

Decigram's mass is not much at all.
A leaf or a feather, she's rather small.

Centigram's mass is so little too.
A few grains of sand
or sugar will do.

Milligram is tiny. POOF! She goes.
She's tricky to measure.
I think she knows.

Now say them in order from greatest of mass.
Read "Gram" in the middle, so you will pass.

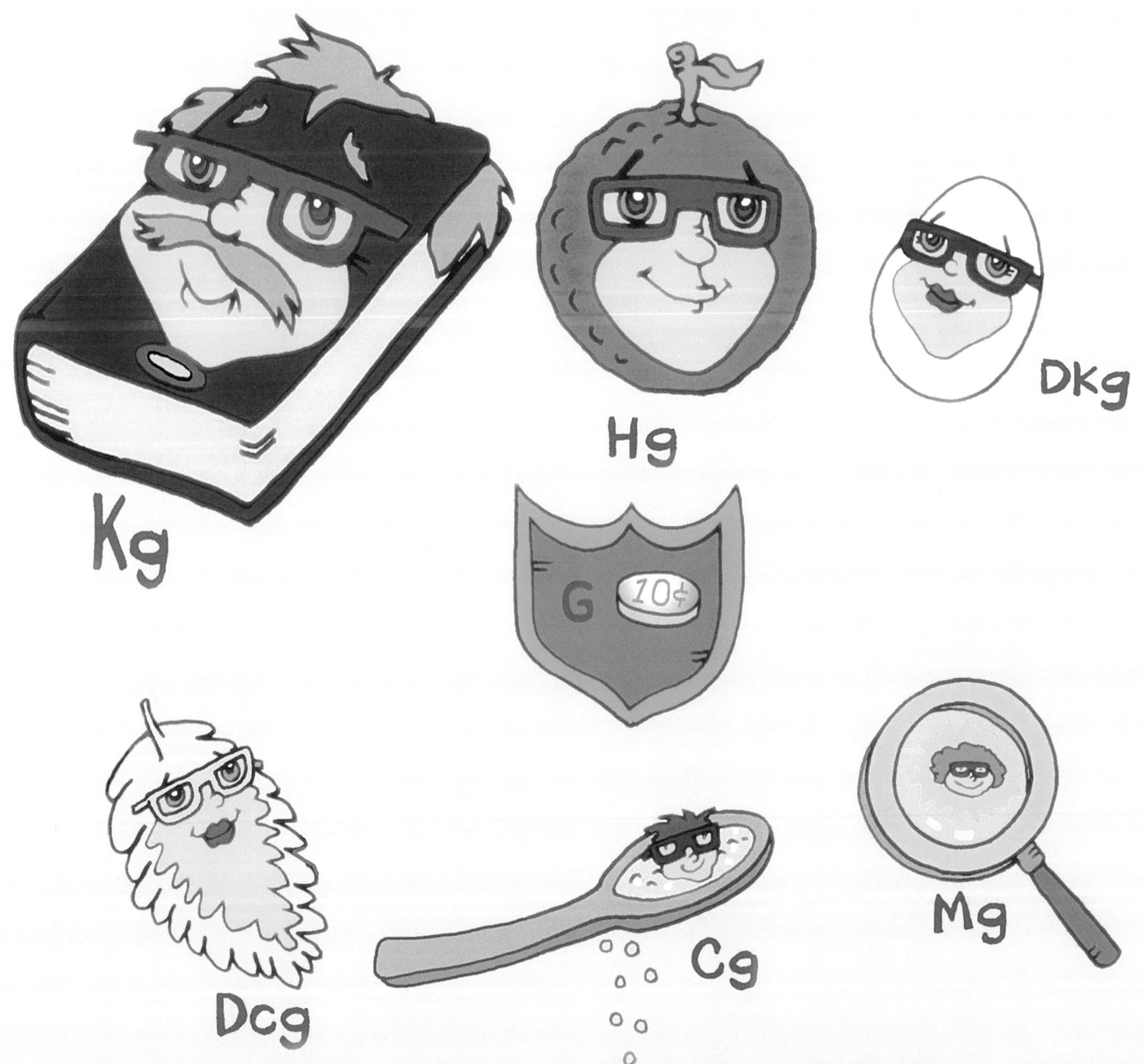

LOOK what has happened to their name!
It morphed to "Meter"...
"Length!" they claim.

When the last name is Meter,
you will know.
They're measuring length or
how far to go.

Grandpa Kilometer is longest by far.
He calculates distance, like in a car.

Hectometer rises very high.
He soars like a building in the sky.

Dekameter stretches...
Watch her grow,
as tall as a tree
from head to toe.

A meter will measure
just shorter than you.
If you were to measure,
you'd find that true.

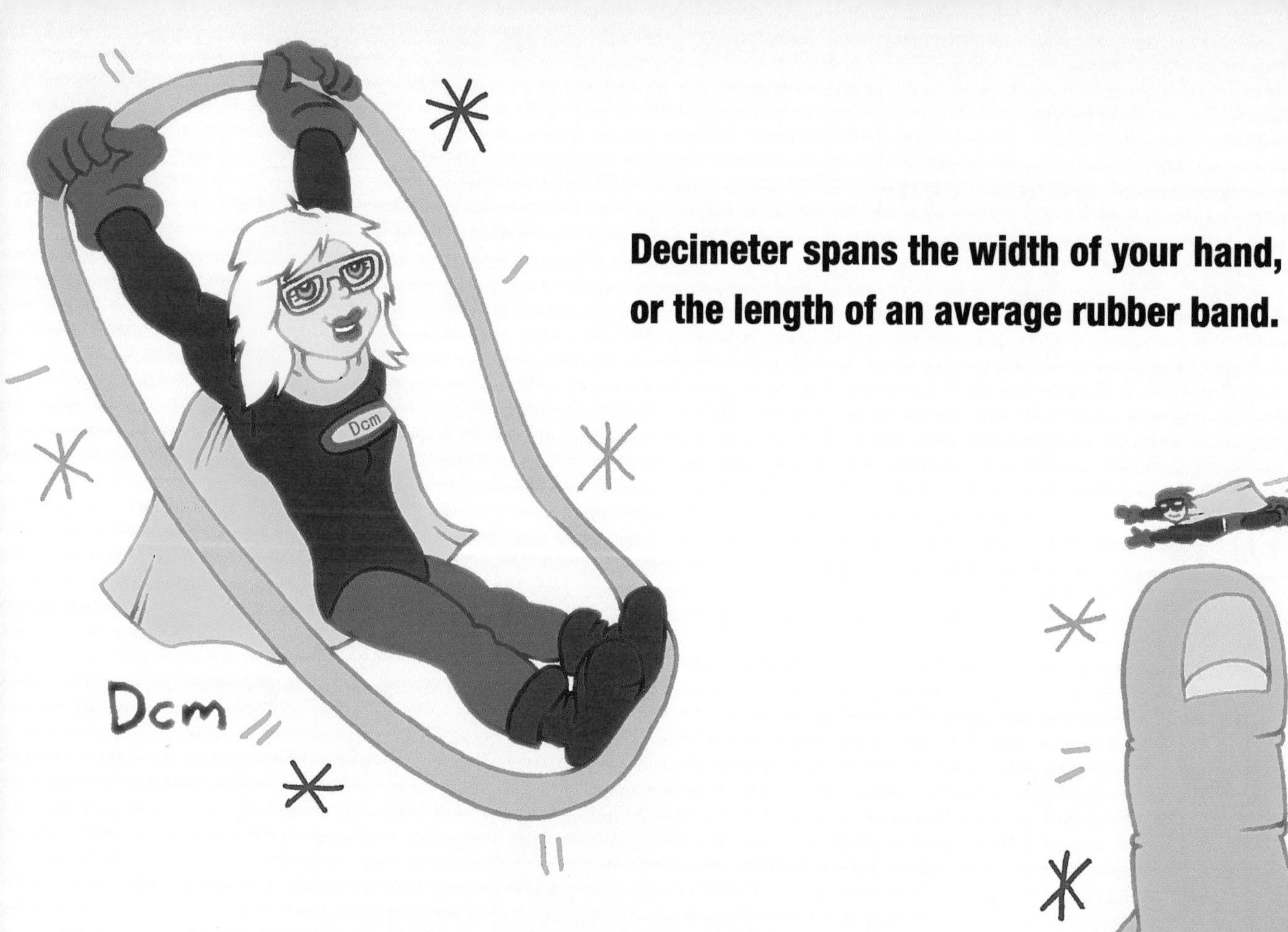

Decimeter spans the width of your hand, or the length of an average rubber band.

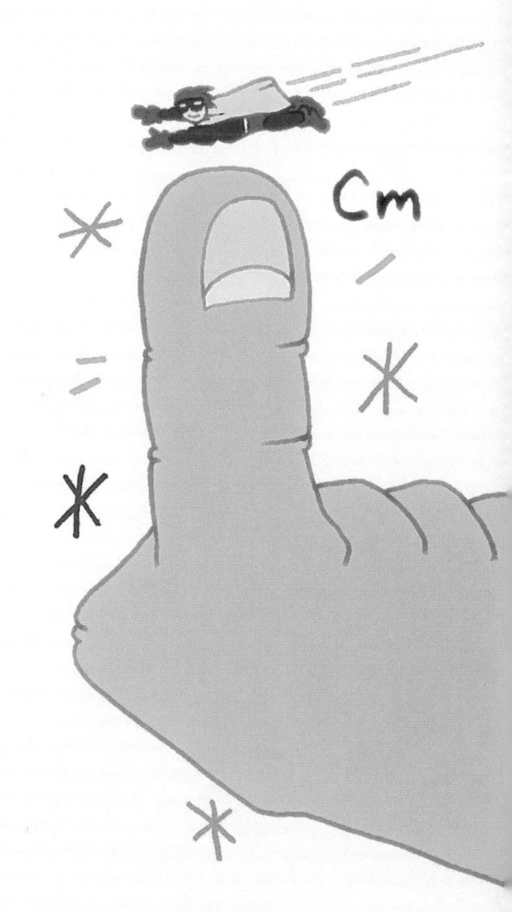

Centimeter is shorter, by ten times as much.
He's the width of your finger,
the end where you touch.

Millimeter is teeny, so very small.
She's the size of a speck, and not very tall.

Longest to shortest, you know how.
The middle is "Meter." Try it now.

The crest changed again!
They have a new name.
It's looking like "Liter" ...
"Volume!" they claim.

When the last name is Liter,
you will see.
They measure volume
or capacity.

To fill a pool, you would ask
Kiloliter to do the task.

When you take a bath and scrub,
have Hectoliter fill the tub.

Dekaliter is often found,
in a bucket, splashing around.

Buy a liter of soda pop,
almost anywhere that you shop.

If you drink a juice box
for breakfast or lunch,
you're swallowing Deciliter,
I have a hunch.

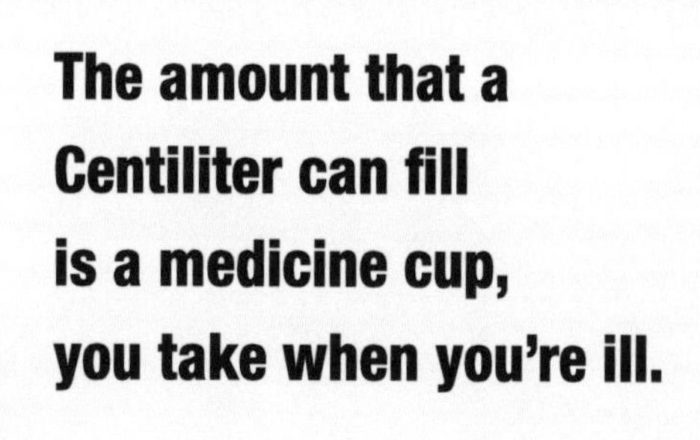

The amount that a
Centiliter can fill
is a medicine cup,
you take when you're ill.

Ml

Watch Milliliter!
Try and stop her.
She can squeeze
out of a dropper.

From greatest to least by volume, please try.
Don't forget "Liter"... You know why.

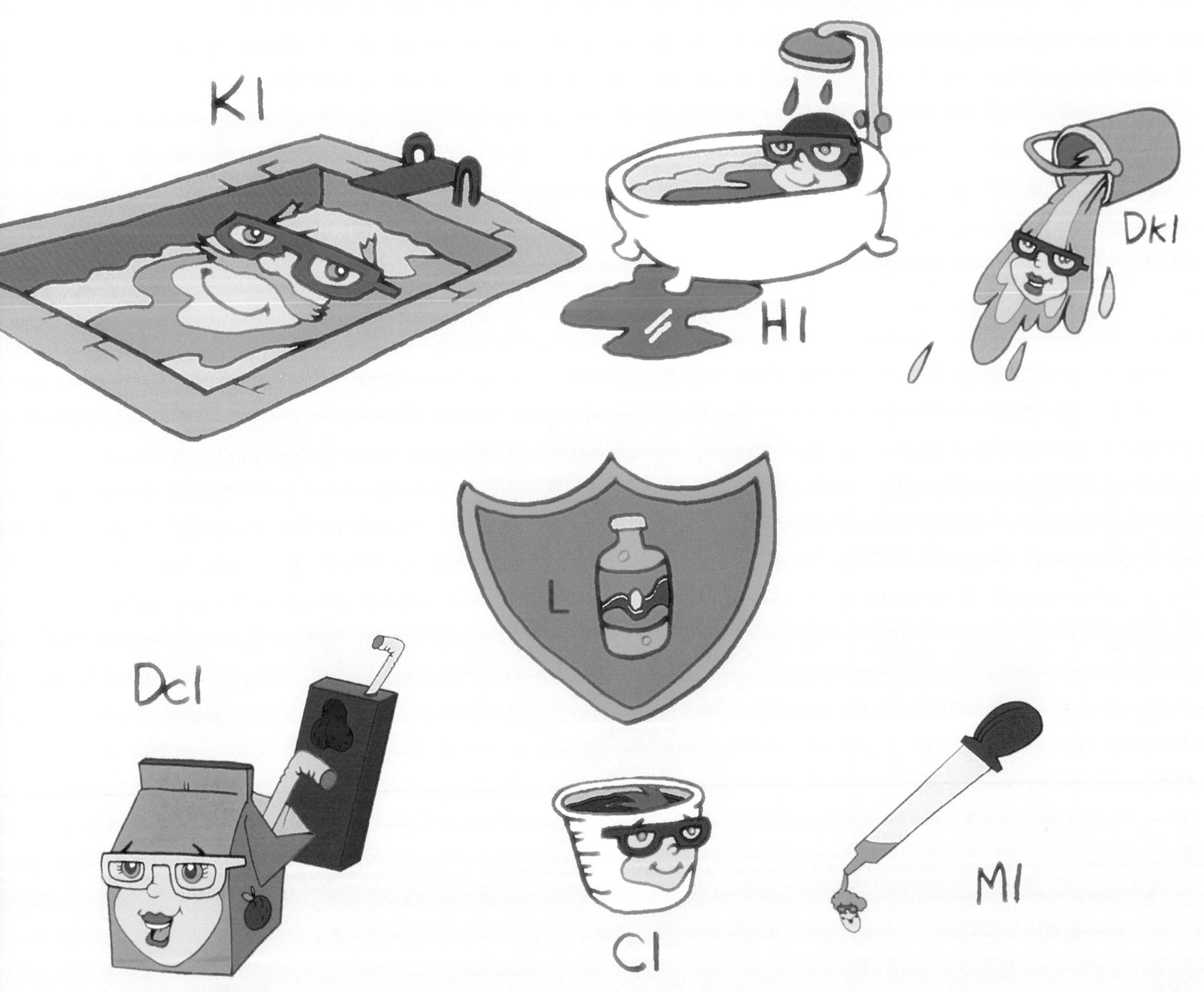

Remember the family … big to small!
Think of base ten as you name them all.

My story has ended, at least for now.
It's YOUR turn to measure! ... YOU know how!

Metric Facts

Divide by ten as the units get smaller in size.
Multiply by ten times as they get larger in size.

GRAM - measures mass
METER - measures length
LITER - measures volume

Fun fact: 1000 kg =1 metric ton

Try These!

Kilo	Hecto	Deka	**Liter** **Meter** **Gram**	Deci	Centi	Milli

1. Which unit would be best to estimate how far it is to the next city?

km   M   kg   cl

2. Estimate the mass of 10 dimes.

1g   1 dkg   1 hg   1 mg

3. If you need to measure food coloring for your cookies, what unit would be best?

kl   L   dcl   ml

4. How many milligrams would it take to make one centigram?

5. How many dekaliters make one kiloliter?

6. How many decimeters are in one hectometer?

Challenge:
How many millimeters are in one kilometer?

(Answers below)

ANSWER KEY

1. km
2. 1 dkg
3. ml
4. 10 milligrams = 1 centigram
5. 100 dekaliters = 1 kiloliter
6. 1000 decimeters = 1 hectometer

Challenge Answer:
There are 1,000,000 millimeters in a kilometer

LaVergne, TN USA
18 January 2011
212977LV00001B

* 9 7 8 1 9 3 5 2 6 8 7 7 2 *